Animal Health
at the
Zoo

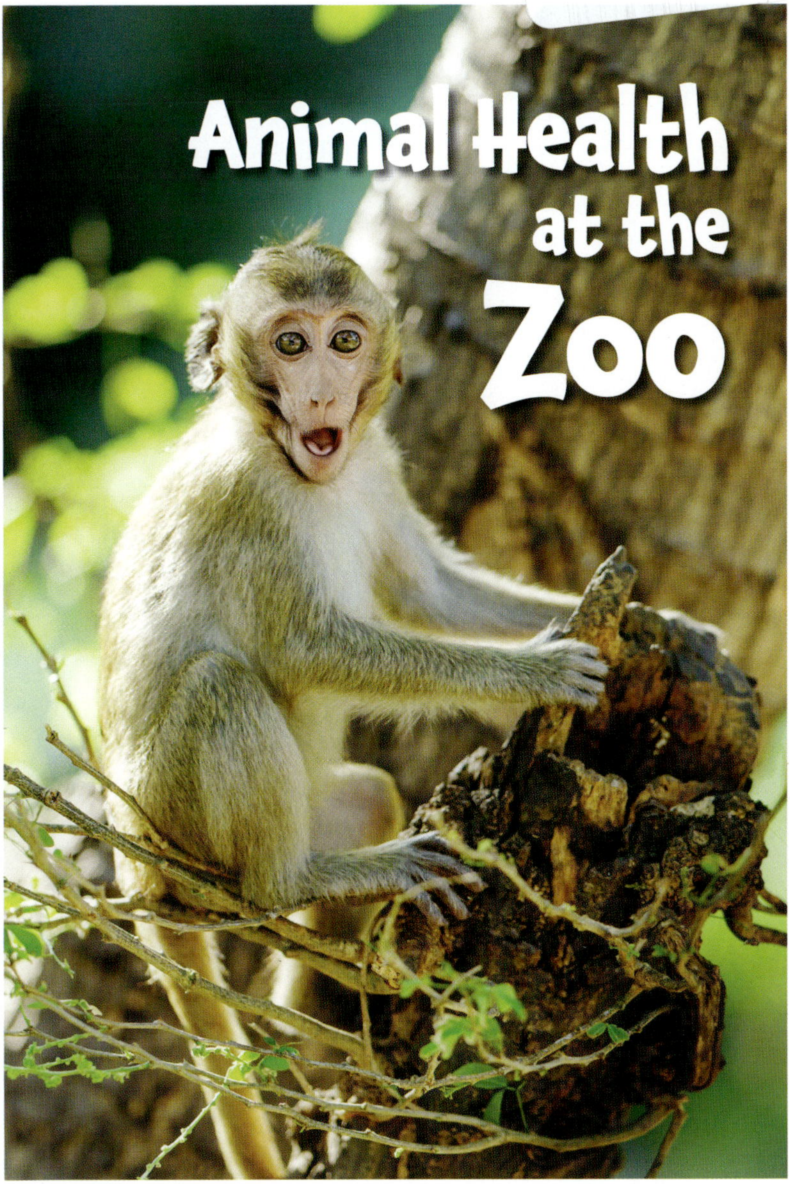

Heather E. Schwartz

Consultants

Jennifer Zoon
Communications Specialist
Smithsonian's National Zoo and
Conservation Biology Institute

Cheryl Lane, M.Ed.
Seventh Grade Science Teacher
Chino Valley Unified School District

Michelle Wertman, M.S.Ed.
Literacy Specialist
New York City Public Schools

Publishing Credits

Rachelle Cracchiolo, M.S.Ed., *Publisher*
Emily R. Smith, M.A.Ed., *SVP of Content Development*
Véronique Bos, *VP of Creative*
Dani Neiley, *Editor*
Robin Erickson, *Senior Art Director*
Jill Malcolm, *Graphic Designer*

Smithsonian Enterprises

Avery Naughton, *Licensing Coordinator*
Paige Towler, *Editorial Lead*
Jill Corcoran, *Senior Director, Licensed Publishing*
Brigid Ferraro, *Vice President of New Business and Licensing*
Carol LeBlanc, *President*

Image Credits: p. 4, Smithsonian Institution; p.9 Smithsonian's
National Zoo; p. 13 (bottom) Adam Mason, Smithsonian's National Zoo;
pp. 15,17, 23, 25 Smithsonian Institution; p. 17 Rotterdam Zoo;
p.20 Gnangarra ;all other images iStock and/or Shutterstock

Library of Congress Cataloging-in-Publication Data
Names: Schwartz, Heather E., author.
Title: Animal health at the zoo / Heather E. Schwartz.
Description: Huntington Beach, CA : Teacher Created Materials, [2024] |
  Includes index. | Audience: Ages 10+ | Summary: ""Zoo animals are living
  creatures who need proper care throughout their lives. Zookeepers work
  to keep animals in zoos safe, happy, and healthy. Encouraging animals to
  satisfy their natural instincts is a key part of a zookeeper's job. Read
  on to learn how animals can live their best lives at zoos!""-- Provided
  by publisher.
Identifiers: LCCN 2024003493 (print) | LCCN 2024003494 (ebook) | ISBN
  9798765968598 (paperback) | ISBN 9798765968673 (ebook)
Subjects: LCSH: Zoo keepers--Juvenile literature. | Zoo keepers--Vocational
  guidance--Juvenile literature. | Zoo animals--Juvenile literature. |
  Zoos--Juvenile literature.
Classification: LCC QL50.5 .S397 2024 (print) | LCC QL50.5 (ebook) | DDC
  590.73--dc23/eng/20240229
LC record available at https://lccn.loc.gov/2024003493
LC ebook record available at https://lccn.loc.gov/2024003494

TCM Teacher Created Materials

5482 Argosy Avenue
Huntington Beach, CA 92649
www.tcmpub.com
ISBN 979-8-7659-6859-8
© 2025 Teacher Created Materials, Inc.
Printed by: 51497
Printed in : China

Table of Contents

Zookeepers Care

Are you an animal lover? Is your idea of a dream job caring for cute, cuddly creatures? Or, do you want to look after unusual or dangerous beasts? If so, you might want to consider a career as a zookeeper.

When animals live at zoos, they are not in their natural **habitats**. These animals can't act exactly how they would in the wild, but they still have natural **instincts**. They were born to behave in certain ways. Zoo animals still need to hunt, eat, rest, and **reproduce**. And, they need to do these things in environments that look like their natural habitats.

Caring for zoo animals is a lot more complicated than looking after domestic pets, such as cats or dogs. For example, hippos need large, clean pools to swim in. Lions need ways to hunt that don't involve killing other animals. Some **species**, including polar bears, may have trouble finding **mates** to reproduce with. If they can't reproduce, their populations could die out. Each species faces different problems and has its own needs.

Zookeepers treat each animal with care.

polar bears at a zoo

This is where zookeepers come in. Zookeepers feed animals, maintain their habitats, and make sure each creature can use its natural instincts. Zookeepers work hard to help the animals in their care live safe, happy, and healthy lives.

Zookeepers help animals in a variety of ways.

5

Lots to Learn

Zookeepers need training to prepare to work with all kinds of animals. In training, they learn skills that will help them do difficult parts of their job. But zookeepers can't learn everything ahead of time. While working, they might experiment with different methods and approaches. They do this because they have to meet the needs of the animals they care for. Zookeepers have to determine what animals need most when it comes to hunting, eating, and reproducing.

An Early Start

People who are interested in zookeeping may start their education in the field when they are young. In elementary and middle school, they may read lots of nonfiction books about animals. They may aim for good grades, especially in mathematics and science. When they get to high school, they can focus even more on these subjects. They can take science classes, such as **biology** and chemistry. Physics and calculus are important as well.

Students who want to become zookeepers can find ways to gain experience with animals. They may work at veterinary offices or animal shelters. Some of them may work on farms. Young adults can also volunteer at zoos. At the Smithsonian's National Zoo and Conservation Biology Institute, students can volunteer when they're 18 years old. They learn about the animals that live there. They also learn how to teach visitors about animal care and **conservation**.

Volunteers are limited in what they can do. They're not ready for direct contact with a zoo's many animals. But volunteering is a great way for people to test out the job of caring for animals. Then, if they like the work, they can take the next steps to pursue a zookeeping career.

Higher Education

Not all zoos require zookeepers to have college degrees. But working toward a degree in animal science, **zoology**, or another life science can be helpful. In college classes, students can learn valuable information about animal care. They typically study higher-level biology and microbiology. These subjects focus on the study of living organisms. Students might also study genetics. This is especially helpful for learning about animal reproduction.

Applying for Jobs

When it's time to find a job at a zoo, applicants with college degrees and relevant experience stand out from the crowd. This is because the field is competitive, even though the day-to-day work isn't glamorous. The job won't lead to great wealth. But people who want to work in zoos are focused and

committed. They're excited about devoting their time and energy to helping animals. And they want to teach others about the animals, too. This is what being a zookeeper is all about!

SCIENCE

Saving the Sick

Zookeepers receive special training on animal behavior. They learn how to look for changes in the ways animals eat, breathe, socialize, and move. These skills are crucial because in the wild, animals hide signs of illness and injuries to stay safe from predators. But in zoos, if animals hide how they're feeling, they won't receive treatment. Zookeepers have to know the signs to watch out for.

Daily Life at a Zoo

Habitats built for zoo animals are made to be **naturalistic**. These habitats mimic the environments where animals would live in the wild. For example, bison have grassy, muddy areas where they can wade and cool off. Crocodiles usually have pools to swim in and dry land where they can rest in the sun. Big cats roam among trees, and they have plenty of logs to use as scratching posts. All habitats are made to help animals feel at home.

Animals living in zoos have limited space. But zookeepers make sure their habitats are clean, safe, and stimulating. Each day, they check on the animals and conduct training sessions with them. They may clean pools or refill holes in yards. Zookeepers regularly add items to make habitats more interesting, too. They might put spices in tubs for lions and tigers to explore. Or they might give gorillas and orangutans a chance to use a touch-screen computer. At a zoo, each day is different!

Every step of the way, zookeepers need to know exactly what they're doing for the animals in their care. Even scooping poop could have unintended consequences. If zookeepers clear away waste too often, they might stop certain species from reproducing. Some animals depend on an odor in waste to know when it's time to seek a mate. So, every action zookeepers take is important.

Elephants need plenty of space in their habitats to roam.

tiger at the National Zoo

FUN FACT

The Smithsonian's National Zoo and Conservation Biology Institute is in Washington, DC. It is home to almost 400 different animal species! The zoo has hundreds of unique habitats. Birds, fish, mammals, reptiles, and insects all need different types of homes.

Dinner Time

Zookeepers and wild animals have one thing in common: they both put a lot of focus on food. Zookeepers need to feed the hungry animals in their care. And wild animals not only need a healthy diet, but a sense of excitement about their next meal as well. Take big cats, for example. As carnivores, they thrive on the ground beef, bones, and frozen meats that zookeepers prepare for them. But their natural instincts can't be satisfied by nutrition alone. They also need to experience the thrill of the hunt.

In zoos, **predators** can't kill natural **prey**. There is simply not enough space for this to happen. Plus, their hunting would harm other animals. To solve this problem, zookeepers often set up safe, **simulated** hunts. They do this for some larger animals, such as lions and tigers.

Lions need to eat fresh meat to survive.

A variety of methods have been used to simulate hunts. Sometimes, toys do the trick. Balls encourage big cats to stalk, chase, and pounce. Other times, zookeepers make the game even more realistic. At the Smithsonian's National Zoo, zookeepers have created cardboard structures that look like prey. Zookeepers can use these to entice animals to act out their hunting behaviors.

Siberian tiger

Safety Systems

Keeping animals safe is important at zoos. But zookeepers have to stay safe as well! All zoos have systems that help keep zookeepers safe. Some exhibits might have color-coded locking systems. Special barriers and railings are typically used. Some exhibits even have special emergency buttons. Zookeepers use them to call for help in dangerous situations.

Some animals don't eat meat, and they like to search for their food. These animals wouldn't be happy or healthy if zookeepers simply delivered their meals. Gibbons, for example, love to find their food. In the wild, they swing through the trees, snacking on fruit. Pandas are another example. They are mostly vegetarian but will eat small rodents occasionally. They enjoy finding bamboo to snack on. To satisfy all these animals' instincts, zookeepers spread their food around their habitats. That way, the animals can **forage** for it as they would in the wild.

white-cheeked gibbon

For enrichment, this zookeeper puts food into a puzzle feeder for the gibbons to find.

Gibbons enjoy eating a mix of fruits and vegetables.

Enrichment

Zookeepers have another way to keep animals' lives interesting. They may offer animals enrichment. This includes providing toys or climbing structures in habitats. Sometimes, they give animals training sessions. New scents, sounds, or changes to habitats are also part of enrichment. All these things make habitats fun and interesting for animals. Enrichment gives animals choices to make and activities to do.

For example, pandas receive a lot of enrichment. Zookeepers put their food inside puzzle feeder toys. Then, zookeepers scatter the toys throughout the pandas' habitat. Pandas get physical exercise when they move around in search of their food. They get mental exercise when they work to solve the puzzles. Their reward is a tasty treat. And they get to feel engaged with their environment, just like they would in their natural habitat!

ARTS

Animal Artists

At some zoos, certain animals are taught to paint as a form of enrichment. Some animals, including lemurs, elephants, and seals, have created paintings. Painting gives them a chance to handle new tools. They can hold the paintbrushes in their hands, paws, trunks, or mouths as they paint across canvases. Plus, they get to spend more time interacting with zookeepers.

Reproduction in Zoos

How do different species survive? They make baby animals! Zookeepers play a role in this process. They know that animals need the right conditions to reproduce. So, zookeepers meet those conditions. For example, they might provide private spaces for giant pandas who won't mate in public view. Taking action like this encourages natural behaviors that lead to mating.

Creating habitats where zoo animals will mate is critical at zoos. This is especially true for **endangered** species. Animals such as Asian elephants, orangutans, and sea lions are rare. So are whale sharks, tigers, and chimpanzees. Each species is at serious risk. They might not survive long into the future without some help.

These Asian elephants were transported to a new zoo to breed with another elephant.

Breeding programs can offer the help animals need. These programs focus on studying animals in zoos and how they mate. Their goal is to increase the population of a species. One example of these programs is the Smithsonian Conservation Biology Institute's Center for Species Survival. Here, scientists spend a lot of time on research. They learn about the kinds of care that different animals need to reproduce. Scientists use what they learn to make decisions that help animals. Monitoring these animals allows scientists to keep rare species safe from **extinction**.

Smithsonian Conservation Biology Institute in Virginia

Clone to the Rescue

Black-footed ferrets are an endangered species. But scientists have been taking steps to stop them from going extinct. Scientists used a special type of technology to increase the population. They took tissue samples from a ferret that died decades ago, and they used the preserved tissue to clone a female ferret. In 2020, the first cloned ferret, Elizabeth Ann, was born!

Asexual Reproduction

Whether in a zoo or in the wild, animals need to carry on their species. Some animals can reproduce *asexually*. This means they don't need a partner to make **offspring**. The offspring that comes from asexual reproduction is identical to its single parent. In other words, the baby animal has the same genetic information. It is a clone. Animals that can reproduce asexually include sea stars, jellyfish, and whiptail lizards.

One example of asexual reproduction came as a surprise to zookeepers. In the early 2000s, a female bonnethead shark lived at the Henry Doorly Zoo in Omaha, Nebraska. She lived in a tank without male sharks for three years. Then, to the shock of the zookeepers, she gave birth to a baby shark. At first, scientists thought that she might have mated with another species. But they ruled that out. They determined the baby was produced through *parthenogenesis* (pawr-thuh-noh-JEH-nuh-suhs). This is a form of asexual reproduction. It happens when an animal cannot find a mate.

bonnethead shark

Parthenogenesis is common in **invertebrates**, such as wasps, bees, and ants. It's common in plants as well. But it's rare in higher **vertebrates**, such as birds, mammals, and reptiles. The instance at Henry Doorly Zoo was the first time scientists saw evidence of this type of reproduction in a shark.

FUN FACT

Komodo dragons are a type of reptile. Normally, they reproduce sexually. But if no males are around, females can reproduce asexually. They lay eggs just as they normally would. Some scientists think that Komodo dragons evolved to have this ability. This is because their natural habitat is isolated. Females may have a hard time finding males.

Sexual Reproduction

In sexual reproduction, a male and female animal mate. That's how they make a baby animal. In this process, baby animals get genetic information from both parents. So, each baby is not identical to either parent. It is not a clone of its parents. Most animals in the world reproduce in this way.

Sexually produced offspring have a combination of **genes**. Genes determine a range of characteristics. Behavior, health, and appearance are all tied to genes. Ultimately, this mix of genes makes each baby animal unique. Each baby animal is different from all other animals within the same species. This means the species has **genetic variation**. This variation is key for a species. A mix of genes makes a species stronger. And genes are passed down as new baby animals are born.

Zookeepers make decisions about which animals should reproduce. They pair up animals so they will mate. They also stop some animals from reproducing. This helps limit the population. It also ensures that there will not be too many animals at a zoo with similar genes. A gene pool without enough variation can make a species weaker over time.

MATHEMATICS

Mix and Match

At zoos, animals live with other animals of the same species. But mating only within the group will not create enough genetic variation. So, zookeepers sometimes match an animal from one zoo with a mate at another. This adds more genes to the mix.

Zookeepers carefully transport animals for breeding.

In the wild and in zoos, rhesus monkeys live in groups with their young.

Raising Young Animals

When animals reproduce, zookeepers watch over both the parents and their babies. Zookeepers have to monitor the instincts that drive wild creatures. Some animals, including birds, might allow their babies to starve. This could happen if their babies are sick or weak. Carnivores, such as lions, might even eat their babies. This only happens in rare cases. The parent might do this if they are hungry and need energy to care for other offspring. In nature, these actions are normal. Animals do these things to help their strongest babies survive. But when animals in zoos threaten their babies, zookeepers step in and take action. They especially have to protect babies that are part of an endangered species. Zookeepers want all baby animals to survive.

Baby tigers drink milk during the early months of their lives.

Baby sloth bears and wallabies can be carried by zookeepers.

Zookeepers are involved in raising young animals in various ways. If a parent is unwilling or unable to care for a baby, zookeepers step in to help raise it. They might mix special formulas for some baby animals to eat. They might feed baby animals, such as tigers, with bottles. Zookeepers can also wear special slings to hold babies the way their parents would carry them in the wild. For example, zookeepers do this for sloth bear babies. At all zoos, raising young animals takes a lot of special care and attention.

A special bottle is used to feed this baby armadillo.

Zookeepers who raise baby animals can't help but bond with them. Animals raised by humans feel the connection, too. Sloth bears might climb on their caretaker's body. A tiger might make happy noises and lick their human parent's hands. A baby chimp might even kiss their caretaker's cheek.

The goal of raising baby animals at a zoo is not to create cuddly, exotic pets. In fact, zookeepers want the animals to get back to living with their own species as soon as possible. This is because some of the skills wild animals need are difficult for humans to teach. Chimpanzees, for example, have to learn how to swing on tree branches. Many animals also need to learn specific social skills so they will fit in with their peers as adults. For example, giraffes and llamas have to learn how to behave in herds. Wolves mate for life and need to know how to behave in pairs and packs. Zookeepers alone cannot teach baby animals those skills.

Zookeepers can bond with the animals in their care.

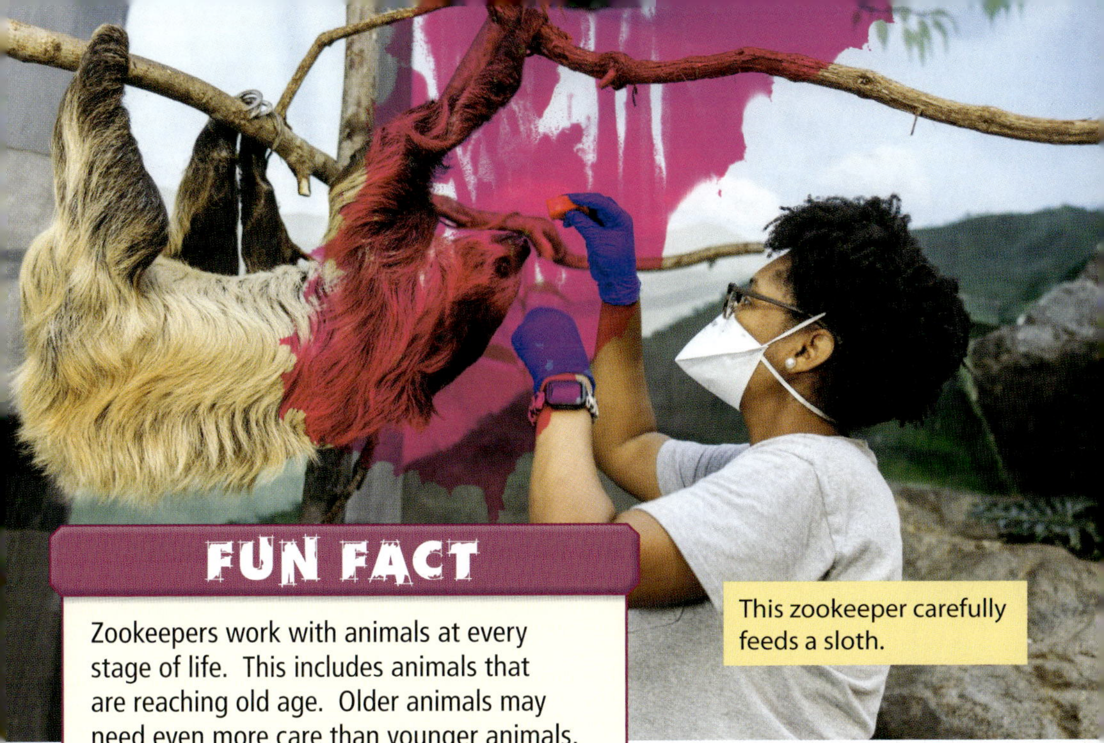

This zookeeper carefully feeds a sloth.

FUN FACT

Zookeepers work with animals at every stage of life. This includes animals that are reaching old age. Older animals may need even more care than younger animals. They may have special diets or go through treatments for illnesses. Some animals can live longer in zoos than in their natural habitats because of the care they receive.

Some studies show that animals who live with their own parents and siblings are better parents themselves later on. But experts also know that certain animals, such as cheetahs, can be fostered by other adults within their own species. As they grow up, they too can learn the parenting skills needed to help the species survive.

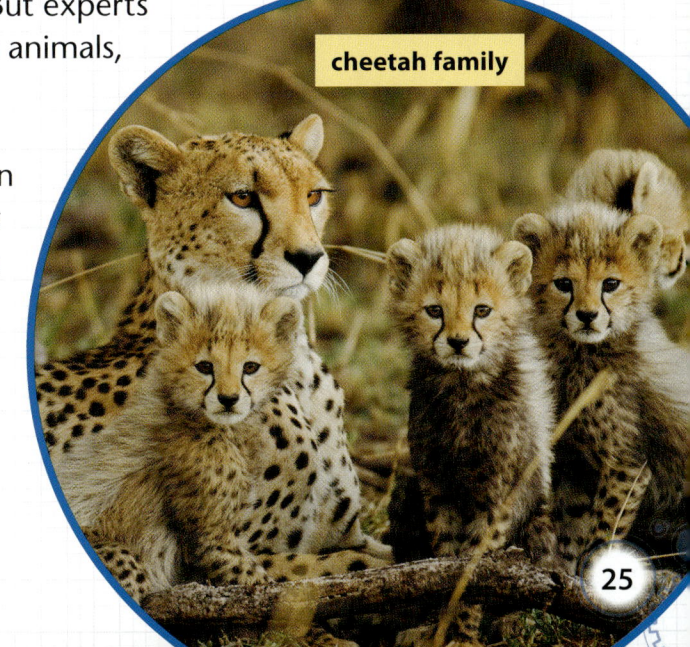

cheetah family

Keeping Animals Safe and Healthy

Zookeepers do everything they can to help animals living at zoos. They feed them, give them water, and monitor their health. They clean their habitats and protect them from danger. They keep their lives in zoos interesting. And they make sure wildlife can behave naturally, as long as their instincts do not hurt humans or other animals.

Whether the animals are babies or adults doesn't matter to zookeepers. It doesn't matter if the animals are cute, cuddly, rough-skinned, or dangerous. Zookeepers always want what's best for all the animals in their care. They work hard to give animals what they need.

Penguins gather around a zookeeper for feeding time.

Zookeepers feed leopards a variety of meat.

In the end, a zookeeper's work is more than a job—it's a mission. Zookeepers care about individual animals. And they think about the big picture as well. They want to help each species survive into the future.

It takes a lot of dedication to become a zookeeper. Zookeepers study animals, understand what they need, and do the physical labor that will help them thrive. Zookeepers are always there for the animals, helping them every step of the way.

This zookeeper helps train a monkey.

STEAM
CHALLENGE

Define the Problem

You are a volunteer at a zoo in your area. The staff at the zoo have asked you to create a new food enrichment toy that can be used in one of the primate (monkey, ape, lemur) enclosures at the zoo. You get to choose which primate you create the enrichment toy for.

Constraints: You may only use the materials provided to you.

Criteria: Your enrichment toy must create a challenge for the animal to get food. Also, it must be appropriate for the primate based on their behaviors, diet, and overall needs.

Research and Brainstorm

What type of primate will you create the food enrichment toy for? What types of food enrichment items are used in zoos for different primates? What do you like and dislike about those designs? What will be challenging or stimulating about the toy you create?

Design and Build

Sketch two or more designs for your food enrichment toy. Label the parts and the materials. Choose the design you think will work best and be most stimulating for the animal. Then, build your enrichment toy.

Test and Improve

Place items in your enrichment toy to represent the animal's food or treats (such as small beads). Show it to others. Explain how it works and why it would be a good source of enrichment for the primate you chose. Then, model how it works. Would it be easy for a zoo employee to prepare? Would it be challenging and stimulating for the animal to use? How can you make it better? Modify your design and rebuild it as needed. Reassess how well it meets the criteria.

Reflect and Share

Was your second design better than your first? How do you know? What surprises or problems did you encounter during this challenge? How did you solve them?

Glossary

biology—a branch of scientific knowledge that deals with living organisms and life processes

conservation—a careful preservation and protection of something

endangered—being or relating to a species that is seriously at risk of extinction

extinction—the process of a species dying out and becoming extinct

forage—search or look for something, especially food or supplies

genes—parts of DNA that pass information from parents to offspring

genetic variation—the presence of different genes in a species

habitats—places where animals normally make their homes

instincts—natural behaviors, abilities, or actions

invertebrates—animals that lack backbones, such as worms, clams, spiders, or butterflies

mates—animals' breeding partners

naturalistic—realistic; natural and lifelike

offspring—a baby animal born from a parent

predators—animals that get food by hunting other animals

prey—animals that are hunted by predators for food

reproduce—to produce offspring

simulated—made to look real but is actually fake

species—a group of living things (especially animals) who can reproduce

vertebrates—animals that have backbones and skeletons, such as mammals

zoology—a branch of biology that focuses on the study of animals

Index

CAREER ADVICE

from Smithsonian

Do you dream of becoming a zookeeper?

Here are some tips to keep in mind for the future.

"Being a zookeeper requires passion. Keepers care deeply about the animals that rely on them for their every need. Those who excel are always educating themselves–through workshops and training–to provide the best care for animals."

– *Jenny Spotten, Asia Trail Animal Keeper, Smithsonian's National Zoo*

"Working with zoo animals has been an amazing opportunity to learn more about their biology and expand my knowledge of animal care. If you have a passion for animals, pursue your passion through volunteering at your local zoo or nature center to better understand what caring for these creatures entails. Be prepared to work hard! Caring for animals is anything but easy, but it's also very rewarding."

– *Jackie Spicer, American Trail Animal Keeper, Smithsonian's National Zoo*